About this Book

If there is one thing that seems unchangeable, it's the calendar. After Monday comes Tuesday. After June comes July. The week has seven days. A year has twelve months. After 1980 comes 1981, etc. But it wasn't always this way. Who said a year has to start here and end there?

By tracking back to find its origins, and then moving forward to find out where and why it kept being changed, the author describes how our calendar evolved and developed bit by bit. In this search we pass through some of the most interesting periods in the history of mankind—the cavemen, the ancient Egyptians, the Romans, the Maya Indians, and others. We learn how each of these different peoples developed a calendar to suit their needs. Step by step, we see how our calendar ultimately came to be the way it is today.

MOON◦MONTHS AND SUN◦DAYS

BY MIRIAM SCHLEIN
ILLUSTRATED BY SHELLY SACKS

YOUNG SCOTT BOOKS

Roman Calender Stone

Published by Young Scott Books, a Division of the
Addison-Wesley Publishing Co., Inc., Reading, Mass. 01867.

Library of Congress Cataloging in Publication Data

Schlein, Miriam.
Moon-months and sun-days.

SUMMARY: Traces the evolution of the modern calendar
from ancient time to the present day.
1. Calendar—Juvenile literature. [1. Calendar]
I. Sacks, Shelly, illus. II. Title.
PZ10.S37Mo   529'.3'09   72-1393
ISBN 0–201–09292–1

Contents

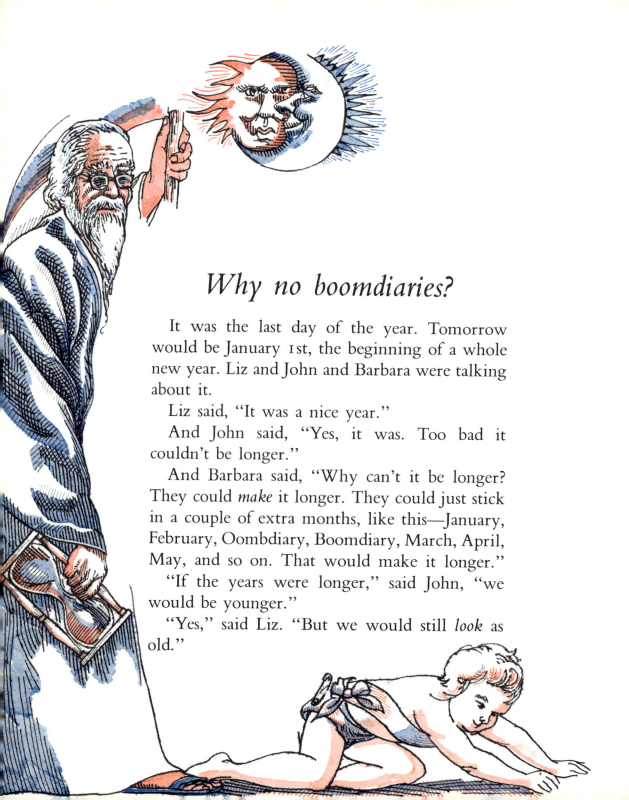

Why no boomdiaries?

It was the last day of the year. Tomorrow would be January 1st, the beginning of a whole new year. Liz and John and Barbara were talking about it.

Liz said, "It was a nice year."

And John said, "Yes, it was. Too bad it couldn't be longer."

And Barbara said, "Why can't it be longer? They could *make* it longer. They could just stick in a couple of extra months, like this—January, February, Oombdiary, Boomdiary, March, April, May, and so on. That would make it longer."

"If the years were longer," said John, "we would be younger."

"Yes," said Liz. "But we would still *look* as old."

"We could have ten days in every week, instead of seven," said John. "That would make the year longer."

"Right!" said Liz. "It could be: Monday, Tuesday, Johnday, Barbday, Lizday, Wednesday, Thursday, Friday, Saturday, Sunday."

They all burst out laughing. They were really just playing with these ideas. Because they knew if there is one thing you can't fool around with and change just like that, it's the calendar. After Monday comes Tuesday; after June comes July. The year 1970 was followed by 1971. The week has seven days, and the year has twelve months. That's something you can count on. You can't go throwing in Oombdiaries and Boomdiaries and Lizdays.

Or could you?

Why can't we make the year longer?

What's a year, anyhow?

Who says it has to start here and end there?

Who decided that?

When?

Why?

Who needs a calendar, anyway?

What's it for?

9

1. The Original Calendar

A calendar is a way of keeping track of time as it passes. It does not have to be a written piece of paper. Primitive people, long ago, did not have a precise, written calendar the way we do, with special names for days and months. But they still could keep track of passing time. Because they had another, wonderful calendar. They had the original calendar—the calendar of the skies. They had the sun and the moon and the stars.

First, there is the sun.
How is the sun a calendar?
The sun gave them the DAY. The DAY is the most obvious unit of time. There is a period of light; then it is over. That is one DAY. They could

see it. It is the easiest way to tell that a certain amount of time has gone by. People could count days, and record their passing.

What makes the DAY?

The earth keeps spinning about. The day-time in any one part of the world is the time when that part is facing toward the sun. The dark-time, or night, is the time when that part of the world is facing away from the sun.

Later on, people began to call the day-time

plus the dark-time one day. This is confusing. Maybe another word should have been made up.

The length of one entire day (one day-time plus one dark-time) is the time it takes for the earth to make one complete spin around on its axis.

Primitive people did not know all this. But they did see there was day-time. And dark-time. It was a good way for them to keep track of passing time.

Then, there is the moon.

How is the moon a calendar?

They looked up at the moon in the sky.

The shape of the moon keeps changing. Or, more precisely, the shape of what we on earth can *see* keeps changing.

Primitive people observed this.

They observed, too, that it always took the same length of time for the moon to go through all its phases—to get from one new moon to the next new moon.

It always took twenty-nine and one-half days.

This gave them a unit of time longer than the day. It gave them a MOON. A MOONTH. We now call it a MONTH.

And so primitive people found that watching the moon was another way to keep track of passing time. They used the moon as a calendar. They could say that a certain trip would take a moon's time. Or they would know from past years that after so many moons of warm weather, they would have to start getting ready for winter.

For how long have people been using the moon as a calendar? We don't know exactly, but it has probably been a very long time. Archeologists recently found some eagle bones and pieces of mammoth ivory that had series of notches and special markings on them. It was determined that these marks were made by prehistoric people—cavemen—who lived in the Ice Age, more than 30,000 years ago. The way the notches are grouped and arranged makes some archeologists believe that the cavemen made these notches to note the phases of the moon and so to keep track of the passing of time, which they may have done as an aid to time their hunting, or for other reasons.

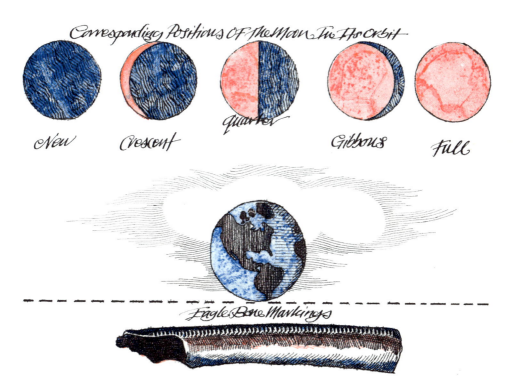

Corresponding Positions Of The Moon In Its Orbit

New Crescent Quarter Gibbous Full

Eagle Bone Markings

Early man had another calendar, too. This one was not in the skies, but on the earth. It was the calendar of nature—of growing, living things. The calendar of animals and plants.

He noticed that certain natural events always happened in the same order. The snow. Then the sprouting of new leaves. The blossoming of certain wildflowers. The ripening of various fruits. The falling of the leaves. Then again, the snow.

In some parts of the world, there was a different calendar of nature. There might be a dry season and a rainy season.

All this took place in the course of one year. Then it started all over again.

What is a YEAR?

While the earth is spinning around on its axis, it also keeps changing its position in the sky, so that it circles around the sun. A YEAR is the length of time it takes for the earth to get around the sun one time. This is called a "solar year." It is 365 days and 5 hours and 48 minutes and 46 seconds long. These changes in the earth's position as it circles the sun during a year is what causes the seasonal changes.

Early man did not know what caused the seasonal changes and events he saw. All he knew was that they did happen. And they happened so predictably each year that he could use their timing as a calendar to help plan ahead. He could, by watching them, reckon the passage of time and plan things accordingly—so that he could be ready for his hunting; so that he could plan his migrations for a time of year that would not be too hot or too cold or too rainy. Later on in man's history, when he planted food as well as hunted for it or gathered it, he could tell, by signs in nature's calendar, when it was the proper time to plant. Perhaps most important of all, the calendar of nature gave him signs each year that the cold season was on the way, and gave him warning to prepare himself for it.

"... When you hear the cry of the cranes going over, that annual voice from high in the clouds, you should take notice and make plans. It is the signal for the beginning of planting.

"... When House-On-Back, the snail, crawls from the ground up the plants, it is time to stop digging in the vineyards.

"... When you first make out on the topmost branches of the fig tree, a leaf as big as the print that a crow makes when he walks, at that time also the sea is navigable."

This advice, based on nature's calendar, was written by Hesiod, a Greek, around 2,600 years ago.

Today, people still pay attention to the calendar of nature. Farmers in northern states say you should not plant corn "until the leaves of the oak are the size of a squirrel's ear."

2. A Special Star

Hesiod also said these things: ". . . When the star Arcturus first begins to rise and shine at the edges of evening, this is the time to prune your vines."

And he said ". . . The morning setting of the Pleiades is the time of sowing and autumn storms."

People have always looked upward to the stars to guide them. They had names for the stars and watched them carefully. For the stars could tell them things. By watching for the rising or setting of certain stars, they could tell that a certain time of year had arrived again.

And so the stars give us still another way to reckon and note the passage of time. The stars, too, are a calendar.

To the people of ancient Egypt, there was one star that was more important than any other. They called it Sothis. It is the star we call Sirius, the Dog Star.

At some times of the year, certain stars are visible to us; at other times they are not. This is because of the movement of the earth around the sun, which enables us to look out on different parts of the heavens at different times of the year.

Each year, the Egyptians would wait eagerly for the day when Sothis would first appear again in the early morning sky. When this happened, it was a time of great rejoicing. For when Sothis was first seen in the sky, it was the sign that the Nile would begin to rise.

Their whole life depended on the Nile. They waited for its waters to rise and flood over the farmlands. When it ebbed, it left the formerly dry, parched earth rich with silt. Then the people could plant and begin the cycle of their life for another year.

And so the rising of Sothis in the summer sky was a great event. And they celebrated it. And the day it was first seen was considered the first day of the new year.

The star Sothis was important to the Egyptians in a different way, too. Their observation of it gave them a new, better calendar.

At first, the early Egyptians had a moon calendar. A lunar calendar. They kept track of time by months. But when they began to celebrate the first appearance of Sothis in the sky, it gave them

a new, measurable unit of time. It gave them the YEAR. From one first appearance of Sothis to the next was a year.

The Egyptian calendar was based on a "star-year" of 365 days. (To be more precise, a "star-year" is 365 days, 6 hours, 9 minutes, and 10 seconds long.) The really proper astronomical name used now for a "star-year" is "sidereal year," and it means the time it takes for the earth to complete one trip around the sun, measured with respect to the stars.

They divided this star-year into twelve months of thirty days each. But this was not enough days to make up a full star-year. So they added five extra days to this. That made a year of 365 days.

The twelve Egyptian months had no names. They were numbered. But the five extra days were named after Egyptian gods—Osiris, Horus, Set, Isis, and Nephythys.

There were three seasons in a year, each four months long.

The first season was Inundation. This was the time when the fields were flooded. (The towns were built up on higher land and so were not flooded.)

The second season was called Seed Time.

The last season was Harvest.

Our own calendar is also 365 days long. But it is based on a SOLAR YEAR, which is also the time it takes for the earth to make one complete trip around the sun. A solar year, however, is measured from one vernal equinox to the next—the time in spring when the center of the sun is directly over the equator, and the day-time is the same length as the night-time.

To be more precise, a SOLAR YEAR is 365 days, 5 hours, 48 minutes, and 46 seconds long. A STAR YEAR, or SIDEREAL YEAR, is 365 days, 6 hours, 9 minutes, and 10 seconds long. Nevertheless, it is said that the Egyptians had the first solar calendar, even though they figured it by means of a star.

This is because a solar year and a star year are in essence very much the same, in that they both include the whole round of the seasons, and are both determined basically by the time it takes for the earth to circle once around the sun.

It is believed that this calendar was begun to be used by the Egyptians in 4241 B.C., about 6,000 years ago.

It might seem that this calendar of 365 days would work out well. There was one problem, though, which was not too obvious at first. But as the years went by, it became more and more obvious.

The civil calendar year was exactly 365 days long. But the star year is not precisely 365 days long. It is about one quarter of a day longer than that. (Six hours, nine minutes, and ten seconds longer.)

So this is what happened. Each year, the man-made civil calendar year completed itself a bit sooner than the "star year" by about one-quarter of a day. So every four years, the *date* of the rising of Sothis would fall one day later in the civil calendar.

Perhaps this doesn't sound so serious. But think of how it was after, say, 400 years had gone by.

The calendar date was then 100 days away from its true seasonal time in the year. The calendar got way ahead of the seasons.

It was silly, and the Egyptians knew it was silly. But they let this error become greater and greater, until what happened? Can you guess? When enough years had gone by, a calendar date would have fallen in every time of year, until finally it once again coincided with the correct seasonal date.

It took 1,460 years for this to happen. They called this a Sothic Period.

Do you know why it took 1,460 years? One quarter of a day $\times$ 1,460 = 365 days. You have

25

gone completely around the calendar. So they just waited 1,460 years, and their calendar was back to normal. Then, of course, the error began all over again.

In 238 B.C., after almost three Sothic Periods had gone by, the ruler Ptolemy III suggested that they should not let this go on. To correct this error, they could simply add one extra day to the civil calendar every four years. This would more or less keep the true season and the calendar date together, where they belong.

Ptolemy III was one of the Greek-speaking rulers from Macedonia that the Egyptians had at this time in their history. Perhaps it is for this reason that the Egyptian priests were stubborn in not wanting to accept his idea for calendar reform. Ptolemy did not insist, and the Egyptian calendar went on as before.

Can you think of a calendar in which Ptolemy III's idea is used?

It is used in our own calendar. Every four years—leap year—we give February one extra day—to keep our man-made civil calendar and the calendar of the seasons going along properly together.

26

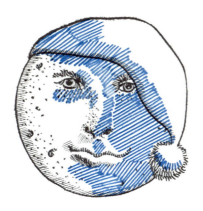

3. Moon-Time

In ancient Rome, they used another sky-calendar to note the passing of time.

They used the moon.

They had officials called pontifices who were in charge of the calendar. Each month the pontifices watched carefully for the first trace of the new moon to appear. When they saw it, they would call out the news. This would tell everyone a new month (moon) had begun. And so the first day of each month was known as the Kalendae, or Callings. (Do you see where we get the word *calendar?*)

The middle of the month—the time of the full moon—they called the Ides. Julius Caesar, the ruler of Rome, is said to have been warned: "Be-

27

ware the Ides of March." The Ides were supposed
to bring bad luck.

The ninth day before the Ides they called the
Nones. (This means nine in Latin.)

If a Roman wanted to indicate a day in between these special times of the month, he would
say, "It is two days before the Ides," or "three
days before the Nones."

The early Romans had an eight-day week. It was not a week with day-names, the way we have. Their week was just the time between one market day and the next.

But the most important unit of time in Rome was the month: the lunar month. A "moon" month. A lunar month is the time between one new moon and the next. It is always $29\frac{1}{2}$ days long. (To be precise, it is 29 days, 12 hours, 44 minutes, and 3 seconds long.)

This early Roman calendar was introduced in 738 B.C. by Romulus, the first ruler of Rome. March 1 was New Year's Day. There were ten months in the year:

Martius (from Mars, the god of war)

Aprilis (perhaps coming from the Latin "to open")

Maius (the goddess of growth)

Junius (after Juno, the goddess of marriage and the family)

Quintilis (meaning fifth month)

Sextilis (sixth month)

September (seventh month)

October (eighth month)

November (ninth month)

December (tenth month)

These months were not divided evenly. Five months had thirty-one days, four months had thirty days, and one month had twenty-nine days.

You may say that this does not add up to a full year's time.

You're right. It doesn't.

But the Romans were rather casual about their calendar. There was a time of year, following December, that they simply didn't care about. It lasted for about sixty days; it was cold, dreary, and nothing much was going on. So they simply left it out of their calendar. This left a big blank spot—a nameless time—each year.

It is thought that it was a later Roman ruler named Numa Pompilius who added two more lunar months, to fill in the blank spot. He called these months Januarius and Februarius (January and February)—Januarius for Janus, the two-

faced god, one face looking back at the old year, one face looking forward to the new year; and Februarius from the Latin "to purify," since the Feast of Purification was held at that time.

The months were redivided so that there were four months with thirty-one days, seven months with twenty-nine days, and one month, Februarius, with twenty-eight days.

Now they had a calendar that added up to 355 days. Or, more exactly, a year of twelve moons—twelve lunar months.

But this still does not equal a true seasonal or solar year. (Remember? It takes 365 days, 5 hours, 48 minutes, and 46 seconds for the earth to get around the sun and go through all the seasons.)

With more than a ten-day difference between the lunar year and the solar year, the calendar date and the season would not move along at an equal pace, and the calendar would soon be in great error.

This would cause difficulties. For example, if the festival Vinalia Rustica, when the priests plucked the first fruits of the vineyard, was celebrated on a certain fixed date in the lunar calendar, each year it would fall eleven days earlier

in the season. When it got too early in the season,
the grapes would not be ripe enough to pick!

This is how the Romans solved the problem.
Every two years, they added a short extra month
of twenty-two or twenty-three days to make up
the difference between the lunar year and the
solar year. Doing this is called "intercalation,"
or inserting something into the existing calendar.
Our leap-year day is an intercalation.

But the Roman pontifices were not exact or conscientious about making this intercalation. Often they would deliberately let the calendar go wrong, just to suit their own or a friend's personal or political convenience. By the time Julius Caesar became ruler, in 49 B.C., the calendar was inaccurate by about two months. The date January 1 came sixty-seven days early. It fell on the seasonal time of October 22.

This was ridiculous. What was the point of having such an inaccurate calendar?

During Julius Caesar's military campaign in Egypt, he had learned about the Egyptians' solar calendar, which, even though it had some faults, seemed much better to him than the lunar calendar used by the Romans. He appointed a Greek astronomer named Sosigenes, who lived in Alexandria, Egypt, to come and study the problem of the Roman calendar, and to suggest ways to improve it.

This is what Sosigenes suggested: that the Romans stop using the moon as a basis for their calendar and use the sun instead; in other words, that they follow a solar year, as the Egyptians did, rather than the lunar year.

He thought the Romans should adopt the

Egyptian calendar system—but with two changes. Instead of having all thirty-day months, with five extra days at the end of the year, as the Egyptians did, he thought that Januarius, Martius, Maius, Quintilis (July), Sextilis (August), October, and December should be given thirty-one days. Februarius was considered an unlucky month and therefore left short, with just twenty-eight days.

The other way he thought that the Roman calendar should depart from the Egyptian was this: he said that every fourth year Februarius should be given one extra day to keep the season and the calendar more precisely in step. This was the idea Ptolemy III had had about 200 years back, which the Egyptians themselves had refused to use, and had still not adopted into their own calendar.

So the idea of leap year, which we use, came first from a ruler of Egypt who himself was not Egyptian but of Macedonian origin; it passed over Egypt, and was first used in ancient Rome.

To start off correctly, and get the calendar date back into its right season, Caesar ordered that that particular calendar year, which was 46 B.C., should have sixty-seven days added to it. Twenty-three additional days had also been intercalated that year. So the year 46 B.C. was a strange year. It had 445 days in it!

The Romans called it "The Year of Confusion." But really, it might have been more fair to call it "The Year to End the Confusion"—the confusion caused by an inaccurate calendar.

A few more small calendar changes were made after this. The Romans changed the name of Quintilis to July, in honor of Julius Caesar. (Quintilis meant fifth month, which it no longer was anyhow, since about 100 years before Caesar's rule, the start of the new year was changed from Martius 1 to Januarius 1, the date that their officials, the consuls, took office.) And later on, Emperor Augustus took the following month, Sextilis (now no longer the "sixth month"), and renamed it for himself—August.

It is basically this calendar, adapted from the ancient Egyptians, suggested by Sosigenes, a Greek, and put into effect by Julius Caesar in Rome, that we use today. It was called the Julian Calendar, after Julius Caesar.

4. A Troublesome Eleven Minutes

You might think that with all this calculating and changing, the Julian Calendar would now work perfectly.

But did it?

No.

There was still a problem.

The addition of one day every four years (leap year) would keep the calendar accurate if the solar year were exactly $365\frac{1}{4}$ days long. Or 365 days and 6 hours.

But it is not. The solar year is only 365 days, 5 hours, 48 minutes, and 46 seconds long. It is shorter than the Julian calendar year by about eleven minutes.

What effect would this difference have?

It means the solar year—or seasonal year—would complete its cycle approximately eleven minutes earlier than the calendar year each year. As many years went by, and all the eleven minutes added up, the solar year seasonal time would fall on an earlier and earlier calendar date.

That is what happened. In the year 1580, the Spring Equinox (the day in which the day-time and the night-time are of exactly equal length) fell on the calendar date of March 11. This is ten days too early. It is supposed to fall on March 21. The civil calendar had fallen ten days behind true seasonal time.

This would not bother the average person. But

it did bother the church officials. They saw that if the calendar date kept drifting this way, religious holidays would begin to fall in the wrong seasons. Christmas would fall later and later in the season, toward spring . . . then summer.

Another big problem was Easter. Easter is not set on a fixed calendar date, as Christmas is. Easter is what is called "a moveable feast." The date for Easter is set each year as the first Sunday after the full moon following the Spring Equinox.

If the solar calendar (seasons) and the civil calendar (date) kept drifting apart, the time of the Spring Equinox would fall on an earlier and earlier calendar date. But according to church custom, Easter was not supposed to fall before April 21.

Again, a calendar correction was needed.

Calendars have always been in the keeping of the priests. This was true of the Hebrews, and the pontifices of Rome, and the astronomer-priests of Egypt.

Now, after studying the problems of the calendar, Pope Gregory XIII ordered that ten days be dropped from October, 1582. This would advance the calendar date, and let it once more catch up to the true seasonal date.

October 5 became October 15. This put the Spring Equinox back on its proper calendar date.

But what would happen in the future? Would not this eleven minute difference cause the calendar to fall behind again?

It would. They thought about it. And this is the way that problem was corrected.

Pope Gregory ordered that three times during every 400 years, the one extra day of leap year would be eliminated. It was to be eliminated in century years that could not be divided by 400. It was eliminated, therefore, in 1700, 1800, and in 1900. In the year 2000, our next century year, Leap Year will *not* be eliminated. It was carefully figured that by means of this adjustment, the calendar year would keep up with the solar year.

Now the calendar was called the "Gregorian Calendar," after Pope Gregory XIII.

Did this last correction finally give us an accurate calendar?

Just about. The Gregorian Calendar is so accurate that the difference between the solar year and the calendar year is only about 26.3 seconds per year.

Can you figure out how many years it would take for the calendar date and the seasonal date to get out of step by just one day, under the Gregorian Calendar?*

Again, it caused some confusion when the new calendar was put into effect. Different countries, for example, adopted this calendar at different times.

Some countries—those that still accepted the Pope's religious leadership—accepted the Gregorian Calendar at once. But this happened to be the time right after the Reformation, a time of great bitterness and conflict between the rebelling new "Protestants" and the Christian Church

* 3,285 years

of Rome, led by the Pope. And so many countries that were mainly Protestant resisted the new calendar, not because it was a bad idea, but simply because it came from a pope, even though it had nothing to do with religion.

England did not adopt the new calendar until 1752. For a while after this, in England, it was confusing. People would refer to a date as being either "old style" or "new style." Old Style (o.s.) meant the date was under the old Julian Calendar that they had been using. New Style (N.S.) meant the date was under the new Gregorian Calendar.

George Washington was really born on February 11, 1731, o.s. But when the calendar change was made, his birthday would have been February 22, 1732, N.S.

You may wonder why there is a year's difference, as well as an eleven-day difference in his birthdate when the calendar had just been changed by ten days.

Under the new Gregorian Calendar, everyone was to observe January 1 as the beginning of the new year. Before 1752, under their old calendar, England had observed March 25 as New Year's Day. Since George Washington's birthday fell in February, this made it one year earlier under the old calendar, which did not begin the new year until March.

The extra day's difference is accounted for in this way: the year 1700 in the new Gregorian Calendar had its leap-year day taken away from it because it was a century year *not* divisible by 400. Since England, in 1700, was still using the Julian Calendar, their calendar had this one extra day in it.

In 1731, Virginia, where Washington was born, was still one of the British colonies, and was therefore also still using the Julian Calendar, the same as England. And so George Washington was officially born on February 11, 1731 (o.s.), and not really on February 22, the date on which we now celebrate his birthday.

5. Some Different Calendars

That is the story of our own calendar, and how it came to be the way it is. It is a collection of ideas from different times and different places and different people. But it was not the only calendar to be developing through all the thousands of years of civilization.

In ancient times, different civilizations were developing in different parts of the world. Sometimes they had contact with one another through wars and invasions, or through trade, or travelers passing through, and in this way ideas were passed on from one to the other.

Sometimes, though, entire civilizations did not even know of the existence of one another. But all of them in their own ways were trying to

grasp the idea of time, and to keep track of its passing. And so they all developed their own different calendars.

Sometimes they borrowed ideas from other societies, and sometimes they did not.

The ancient Greeks had a solar-lunar calendar. Their figuring of a month was based on the moon. Their figuring of a year was based on the sun.

None of the ancient peoples knew what actually caused a solar year. They did not know that the earth circled the sun, and that this took 365 days, 5 hours, 48 minutes, and 46 seconds, and that this is what made a "solar year." They figured their "year" from one solstice to another solstice. Or from one equinox to another equinox. (The winter solstice is the shortest day of the year; the summer solstice is the longest. The two equinoxes, in autumn and spring, are the days on which the light-time and the dark-time are of equal length.)

The Greek calendar had "oktaeteris." These were groups of eight years.

Eight solar years are 2,922 days.

Eight lunar years (eight years of twelve lunar months each) are 2,832 days.

There is a difference of ninety days. So, at three different times during the eight-year group, the Greeks would intercalate one month, so as to keep the eight-solar-year group and the eight-lunar-year group in pace with one another, and also to keep the calendar date and the seasonal time in pace.

Later on, in 432 B.C., a Greek named Meton, who lived in Athens, figured out a system that

was more accurate. This was to have groups of nineteen years, instead of eight years. These were called "great years." Into these great years they intercalated seven months, at various times.

The Hebrew calendar used by the Jews started in the year 3760 B.C., which was, they believed, the date of the Creation.

It was also a lunar-solar calendar, with twelve lunar months in the year. Each month had twenty-nine or thirty days.

In earliest times, the beginning of each month

was set by direct observation of the moon. Witnesses who first saw the new crescent in the sky would come forth and report it. They were carefully interviewed, and if their information was confirmed by astronomical calculations, the Sanhedrin, the Supreme Court in Jerusalem, would announce the beginning of the official new month. Fires were then lighted on the mountaintops at night to signal to Jewish communities that the new month had begun. In later times, messengers were sent out with the news.

However, they did also keep this lunar calendar in time with the solar year, which has in it about eleven days more than a twelve-month lunar year. They did it by adding in, or intercalating, extra months.

The Calendar Council of the Sanhedrin was in charge of intercalation. When the lunar calendar was about thirty days ahead of the solar calendar, they would add to the year a thirteenth month called Adar II before the month of Nisan, when Passover occurred. In this way they would ensure that Passover would not fall too soon in winter, rather than spring, when it was supposed to be celebrated.

The Calendar Council decided when intercalation was needed by means of astronomical figures which had been handed down to them. Also, they used the calendar of nature to tell them when an additional month was needed to keep the lunar calendar in step with the solar year. It is written in the Talmud, the book of Jewish laws and customs, that the Council would intercalate a month if by a certain date the barley in the fields had not yet ripened, the fruit on the trees had not grown properly, the winter rains had not stopped, the roads used by Passover pilgrims had not yet dried up, and the young pigeons had not become fledged.

All these things were signs in nature's calendar.

This system was used until the fourth century A.D., when the very existence of the Jewish Sanhedrin was threatened by unrest and persecution. If there were no Sanhedrin in Jerusalem, and no Calendar Council to give them guidance, how would the Jews scattered all over the world keep a unified calendar and know when to celebrate their important holidays?

It was for this reason that Rabbi Hillel II publicly announced the system of calendar calculations which, up until that time, had only been

known to the Calendar Council. Now these calculations could be used by Jews all over the world to keep their calendars accurate and similar.

In the year 358 A.D. Rabbi Hillel II adapted the idea of the Greek nineteen-year Metonic cycle to the Jewish calendar, which provided a regular, consistent form of intercalation.

This is the way it worked. In each nineteen-year span, there were seven special years—the third, the sixth, the eighth, the eleventh, the fourteenth, the seventeenth, and the nineteenth. To each of these years, an additional month was added. This kept the lunar calendar accurately in pace with the solar year.

The Hebrew calendar has continued, and is in use at the present time. Jewish holidays are still set and observed all over the world according to the Hebrew calendar.

Our year 1971 was 5731 in the Hebrew calendar.

At around 3000 B.C., the ancient Sumerians had a calendar based on the moon. They had twelve lunar months, with thirty days in each month. Their months were named after events

A Sumerian Moon Disc 4,600 years old
A priest sacrificing to Nin-Gal, wife of the
Moon God

in their daily life. Some of these were: Month-of-the-Leading-Out-of-the-Oxen, Month-of-Brick-Making, Month-of-Opening-the-Irrigation-Canals, Month-of-Ploughing, Month-of-Corn-Harvest.

Here again we find a calendar year based on moon-months, which is not as long as a solar or seasonal year. But we do not know exactly if or how the Sumerians intercalated, or made their calendar "corrections."

In nearby Babylonia, from around 2000 B.C., they also had a year of twelve lunar months. But here we know that they did make calendar corrections. They simply added a short intercalary month when it was necessary—that is, when it was noticed by the priests, who would then remind the king "that the year hath a deficiency."

Later on, the Babylonian astronomers discovered what is called a "phase-relationship" between the sun and the moon every nineteen years. They saw that nineteen solar years were almost exactly as long as 235 lunar months. 235 lunar months is what you get if you take nineteen years of twelve lunar months each, and add seven months to it. So, from 382 B.C, the Babylonians started adding, or intercalating, the needed months in a regular, organized way. They added seven extra months during the nineteen-year period.

This nineteen-year cycle is called the Metonic cycle after Meton, the Greek, who started its use in Greece in 432 B.C.

In another part of the world, at about the same time, there was a group of people who had no contact at all with any of these other civilizations and their calendars, but who developed, by themselves, a calendar that was not only different from all of the others, but was amazing in its accuracy. These were the Maya Indians, who lived in Central America. Their civilization began at an earlier date, but it was during the time from about 357 B.C. to 317 A.D. that they developed a calendar which was, even then, more accurate than the one we use today.

The Mayas had a number system based on twenty. They counted fingers and toes. (Ours is based on ten. Just fingers.) Their calendars were based on very exact observations the priests made of the sun, the moon, and the planet Venus.

Their civil calendar was based on the solar year. It had eighteen months, with twenty days in each, and a nineteenth month with five days in it.

This adds up to 365 days, which would bring to the Maya civil calendar the difficulty we saw in all the other calendars. Since 365 days is nearly six hours less than the true solar year, this would cause any date to fall earlier and earlier in the true season with every year that went by.

But the Mayas, unlike other people, did not correct their calendar by intercalation. This is why. The Mayas had other calendars, in addition

Mayan Calendar Wheel

to their civil calendar. They had a lunar calendar, they had a Venus calendar, and they also had a Sacred calendar.

All of these calendars were arranged to go along in relationship to one another. Like gears of different sizes, the calendars of different lengths were linked to one another. So, repeatedly, every certain number of years, the different calendars would come into the same relationship with one another. Once every fifty-two years, for instance, the day in a certain position in the week would fall on the same date. This was called the Calendar Round.

If intercalation were to be made on the civil calendar, it would destroy this relationship. So, instead of intercalating, the Mayas would simply change the calendar dates of their holidays and farming activities when they began to fall into the wrong month.

Their Sacred calendar had 260 days. It was divided into thirteen-day periods like weeks, and it had twenty different day-names running through these thirteen-day weeks. (It was as though we had a seven-day week, but had ten day-names. All the day-names would not be used up in the first week, and would be used for the first time in the second week.)

A period was considered lucky or unlucky depending on the day-sign that it began with. People born in a week starting with Cipactli (the mythical water monster) would be happy and fortunate. But people born in a week starting with Eecatl (the wind) would turn out to be traitors and witches and have the power to turn themselves into animals. Each period beginning with each different day-sign had a different prediction for people born during that time.

6. *The Week*

All these calendars—the Egyptian, the Hebrew, the Roman, and the others—are different from each other in many ways. But in some ways, they are very much alike. They are alike in this way: every calendar that man has invented to indicate the passing of time has been based on the calendar of the skies—the relationships and movement of the earth, and the sun, and the moon, and the stars. Their appearance, their reappearance, their position in the sky. These movements and relationships are what create natural units of time. The day. The month. The year. And so, these same units appear in just about every calendar.

The WEEK is something else. The week is not a natural division of time, based on nature. It is an artificial length of time, made up by man.

Where did the idea of a WEEK come from?

The idea of a week began simply as the time between one market day and the next. People in different societies had weeks of from four to ten days, depending upon how often they had market days.

The ancient Greeks had three ten-day weeks in their month. The Egyptians also had three ten-day weeks in their month. They were called "decads."

Why does our week have seven days?

The seven-day week goes back a long way. The Babylonians and Assyrians had seven-day weeks as far back as 2300 B.C. The Chaldeans, who lived in the same area, believed that the seventh, the fourteenth, the twenty-first and the twenty-eighth of each month was a day of bad luck. They therefore held back from doing important things on those days.

Some think that the seven-day period was suggested by the phases of the moon, each of which lasts about seven days. Or, perhaps it was the fact that seven was often thought of as a sort of magic number.

The Babylonians divided time into seven-day

weeks to honor the seven heavenly bodies which they worshiped as gods—the sun, the moon, and the five planets that were visible to them—Saturn, Mars, Mercury, Jupiter, and Venus.

But it was the ancient Jews for whom the seven-day week became an important element of time in a religious sense. Moses, who led the Jews out of slavery in Egypt around 1400 B.C., introduced the idea of a regular seven-day week, with

the seventh day treated as a holy day; a day of rest and prayer. And this idea was written into the Bible—the Old Testament—by Moses, in the section called Exodus.

"Remember the sabbath day, to keep it holy. Six days shalt thou labour, and do all thy work; but the seventh day is a sabbath unto the Lord thy God; in it thou shalt not do any manner of work . . ."

It was in this way that the idea of the seven-day week became reinforced, and came to be passed down to us.

More than 1,500 years later, Constantine the Great, Emperor of Rome, adopted this seven-day week for the Romans.

SUNDAY   MONDAY   TUESDAY   WEDNESDAY
THURSDAY   FRIDAY   SATURDAY   SUNDAY

Where did we get the names for our days of the week?

SUNDAY is the day sacred to the sun.

MONDAY is the moon's day.

Four of our day-names come from the names of Norse gods worshiped by the Saxons, who invaded Britain about 1,500 years ago:

62

TUESDAY is Tiw's day, from Tyr, the Norse god of war.

WEDNESDAY is Wodin's day, the chief god in Norse mythology.

THURSDAY is Thor's day, the Norse god of thunder.

FRIDAY is Frigg's day, the Norse goddess of love.

SATURDAY is different. SATURDAY, the seventh day, is named for Saturn, the Roman god of the harvest. Many English words came originally from the Roman language, Latin, from the time when the Romans invaded Britain, in 43 A.D.

More calendar changes?

Liz and John and Barbara are not the only ones who have wanted to change the calendar. Other people have wanted to change it, too. Not because they wanted to keep a nice year going on longer, but because they felt that our calendar is not logical. The weeks have no relationship to the months. And the months are of different lengths.

Different suggestions for changing the calendar have been made by different people. One of them is this: that we should have a thirteen-month calendar. We could then have exactly four weeks in each month. The month to be added would be called Sol and come just before July.

A Year Day (to give us our 365 days) would come at the end of the year, and belong to no month. In leap year, leap-year day would be added just before July.

Each year would begin on a Sunday and end on a Saturday. Each date would therefore always fall on the same day of the week.

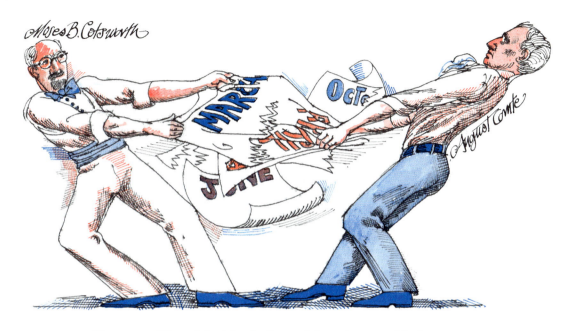

Do you like this idea? One thing it would do would be to give us an extra summer month. Also, it would be easier to remember the date. For example, the first, the eighth, the fifteenth and the twenty-second of the month would always fall on a Sunday. Other dates would be similarly constant.

This idea is based on the thirteen-month calendar suggested in 1849 by Auguste Comte, a French philosopher. Changes in it were made by Moses Cotsworth, a Canadian. It was called the International Fixed Calendar.

There are some religious groups that do not want this or any other calendar changes. They say that the extra days, not a part of any week, would destroy the seven-day cycle of six work days followed by a Sabbath, and so make it impossible to observe the idea stated in the Bible: On the seventh day . . . thou shalt not do any work.

Besides, most people are not at all upset about the calendar as it is now. We accept it with its peculiarities. And so we will probably keep going the way we have been:

Thirty days hath September,
April, June, and November;
All the rest have thirty-one,
Except the second month alone,
To which we twenty-eight assign,
Till Leap Year gives it twenty-nine.

If a Maya Indian or an ancient Egyptian could see our calendar, maybe he or she would think: "That certainly is a strange calendar!"

And if Liz and John and Barbara still want to try to change the calendar, they know now that they can't just do it any old way. They know that there are many considerations they must keep in mind in order to keep the date in season, and to keep everything going correctly year after year.

It's not an easy job. But they are all smart. And so maybe they still want to try to figure out some new calendar ideas.

Index

About the Author

MIRIAM SCHLEIN was born and grew up in Brooklyn, New York, attending Brooklyn College where she received a B.A. degree in Psychology and English.

Author of forty-four books for children, her first, *The Four Little Foxes*, was published by William R. Scott in 1952. A number of her books have been Junior Literary Guild selections and several have been translated into different languages. One book, *Little Rednose*, was an Honor Book in an Herald Tribune Spring Book Festival.

She has had adult fiction published in *Redbook, Ladies Home Journal, McCall's, The Colorado Quarterly,* and *The University Review.*

She lives in Westport, Connecticut, with her two children, Elizabeth and John, and spends her summers on the island of Martha's Vineyard.

About the Artist

SHELLY SACKS was born in Brooklyn, New York. After graduating from art school, he began his career by serving an apprenticeship in an art studio.

He has a wide range of interests extending from collecting old fountain pens and seed catalogs, to opera, cooking, and gardening.

He is a member of a gourmet club whose members meet once a month to cook specialties for one another. It seems appropriate, therefore, that his illustrations decorate a half-dozen cookbooks. In addition, he has illustrated several juvenile books and done a variety of advertising work.

Mr. Sacks and his wife and their three children live in Oakhurst, New Jersey.